Gavin the Gator
Greater Than and Less Than

Kathleen L. Stone

Copyright © 2016 Kathleen L. Stone

All rights reserved.

ISBN–13: 978-1539166672
ISBN-10: 1539166678

Enjoy these other books by Kathleen L. Stone

Penguin Place Value
A Math Adventure

Number Line Fun
Solving Number Mysteries

Riley the Robot
An Input/Output Machine

Mason the Magician
Hundreds Chart Addition

Katelyn's Fair Share Picnic
More Math Fun

Money Tree Mysteries
Adventures with Quarters

Alien Even and Alien Odd
A Math Space Adventure

Kenley's Line Plot Graph
Another Math Adventure

Matthew's Sunshine Bakery
Multiplication Arrays

Firefighter Gary's Fire Safety Rules

Samantha's Search
3D Shapes

Grandma's Quilts
Fun with Fractions

Daniel's Day of Multiplication
Multiplication with Equal Groups

More Penguin Place Value
Hundreds, Tens, and Ones

Tick Tock Telling Time
Time to the Hour and Half Hour

Dedication

Thank you to all the math teachers out there who help children improve their skills and learn to love math!

Gavin the Gator
Loves to spend his days
Swimming in a muddy swamp
Looking for ways
To fill his empty tummy.
He's as hungry as can be.
But being a picky eater,
He won't eat just anything he happens to see.

Snakes, fish, birds, and frogs
Would all make a tasty treat.
But he passes them all by.
Numbers are all he wants to eat.

Swimming through the swamp,
He's a skilled navigator.
And only makes a meal
Of the numbers that are greater.

Early in the morning,
Before the sun begins to shine,
Gavin finds two numbers,
Twenty-three and *fifty-nine*.

With a twinkle in his eye
And a smile of glee,
Gavin gobbles *fifty-nine*
Because *fifty-nine is greater than twenty-three.*

Gliding through the swamp
In an early morning fog
Gavin spies two numbers
Resting on a log.

Which will make a tasty breakfast?
SMACK … SNAP … CHEW
In an instant *seventy-four* is gone
Because *seventy-four is greater that sixty-two.*

Gavin sees a mid-morning snack
Hiding in some sticks.
Which one will he eat?
Seventeen or *thirty-six*?

He swims past *seventeen*.
This sly gator is full of tricks.
He's not wasting any time
He knows, *seventeen is less than thirty-six*!

Searching for his lunch
Under the hot, steamy sun
Gavin spots two more numbers
Forty-nine and *eighty-one*.

On which of these numbers
Do you think Gavin will dine?
He's already made his choice
Because *eighty-one is greater than forty-nine.*

Nighttime is coming.
Soon his day will be done.
For dinner should he choose
Nineteen or *ninety-one*.

The choice is easy for Gavin.
In fact it's kind of fun
Because just like Gavin, I bet you know
That *nineteen is less than ninety-one.*

So if you're ever in the swamp
Be careful who you meet.
It might be Gavin the Gator
Hunting for a tasty treat!

Number and Operations … Comparing Numbers … Greater Than and Less Than

Developing good number sense, recognizing the quantity represented by numerals and how numbers relate to other numbers, is an important skill for young children. Number sense develops over the years as children explore numbers, visualize them in different contexts, and connect to them in a variety of ways. The concepts of "greater than" and "less than" are important skills, not only in real life (Should I hire the painter who charges $15 an hour or the one who charges $25 an hour?) but in other math skills as well.

If a child is having difficulties recognizing which number is greater, use objects to help them visualize the concept. Poker chips, cubes, dried pasta or beans can all be used for children to count out the quantities to compare. The following activities will provide additional practice using the "greater than" and "less than" vocabulary.

Enrichment Activities

Greater Gator Roll

Materials needed:

Pair of dice or number cubes
White boards (or laminated piece of construction paper)
One dry erase pen for each student

Before each game, partners decide if the winner will have the number that is greater or smaller.

How to Play

Children take turns rolling the dice and deciding what number they will make from the numbers rolled. For example, if the child rolls a "two" and a "five" they can either make "twenty-five" or "fifty-two." After both numbers have been recorded on their white boards, they decide which symbol (>, <, or =) to write to make a true equation.

Hungry, Hungry Gators

Materials needed:

Number Cards (you can determine the level of difficulty)
Paper plates

Have children color their paper plates green on both sides and then cut a triangular shape to form a "mouth." Have them use their scraps to make "teeth" for their hungry gators. Don't forget to add eyes.

Place numbers around the room. Working with a partner, children can go around the room and use their hungry gators to make true equations.

92   56

You could also use popsicle/craft sticks, painted green, and googly eyes to make your hungry gators.

ABOUT THE AUTHOR

Kathleen Stone is a National Board Certified educator and is currently teaching second grade. *Gavin the Gator* is her sixteenth children's book. Born and raised in Washington State, she and her husband Gary live in the Olympia area. When not teaching, Kathleen can often be found quilting, sitting by the lake reading, and enjoying time with her family (especially her grandchildren)!

*Math is all around us
No matter where you turn
Open your mind to the wonders of math
And all that you can learn*

Made in the USA
Columbia, SC
09 May 2023